Section 1
Note taking

THIS SECTION GIVES TIPS ON PRODUCING A CONCISE SET OF NOTES FOR FUTURE REFERENCE

Notes are taken at research seminars, lectures and meetings. These notes are often used for revision, to follow up a research topic, or to produce accurate minutes of a meeting.

It is difficult to take notes when there is a lot of information to sort through quickly (both verbal and visual) and you are trying to understand the topic at the same time – sometimes lecturers forget that you are not familiar with the subject they are discussing. Presenters have diverse styles and their handouts vary greatly.

It is useful to summarise information as notes when revising for examinations.

Research seminars

Research seminars and one-off lectures vary hugely in style and often contain lots of information. They can provide information on current research and it is important to attend seminars regularly so you can keep up to date. Your notes should provide an overview of the seminar and include relevant schemes.

Tips

- **Use a hardback notebook to record your notes;**
- **include schemes and diagrams; and**
- **note any references given.**

Lectures

To get the most out of lectures, try to prepare reasonable notes first time, and then go back over them later to check that they make sense.

It is important that your lecture notes are adequate, as they provide the basis of your course syllabus and revision for examinations.

Tips

- **Write headings for each part of the lecture (highlight, underline, and/or number them);**
- **check you have all the important information in your notes or on a hand-out;**
- **use different colours for the main part of your notes and for extra information or to clarify schemes;**
- **use a margin to highlight things, for example:**
  - *** for important points**
  - **† for incomplete notes**
  - **? for parts you didn't understand**
- **develop your own style and abbreviations;**
- **use bullet points; and**
- **include structures, schemes and graphs.**

With practice you should be able to get adequate notes first time and just follow up your margin symbols after the lecture.
It is important to listen and try to understand rather than just copying.

Summarising information

When preparing reports and presentations, or revising for examinations, you may have to extract information from textbooks or journals. You may also be required to produce notes of your own presentations to give to colleagues. Remember to write your notes in a way that others will understand them.

Tips

- **Jot down keywords and relevant structures;**
- **add some references and any additional information so the notes can be understood;**
- **for presentations, reproduce overheads/slides on a reduced scale and include keywords and references alongside each one.**

It is also useful to take notes during meetings - see Section 8 for more tips.

Remember to check your notes and follow up anything you don't understand soon after the lecture or seminar.

Section 2
Keeping a laboratory notebook

THIS SECTION GIVES SOME GENERAL HINTS ON PRODUCING CLEAR, CONCISE LABORATORY RECORDS

Keeping laboratory notebooks is essential for successful scientific research. They provide an accurate record of what happens in the laboratory and are the basis for reports and publications. Laboratory records can be used in legal cases – e.g. proving a patent or if an accident occurs.

Laboratory work is divided into three types: following known procedures; modifying known procedures; or carrying out new research. Each organisation has its own specific requirements for writing laboratory records. Enough information should be included so that a colleague can repeat the experiment using your laboratory notes.

Make sure you are familiar with any specific requirements and adhere to them. This is especially important in industry where you need to comply with good laboratory practice (GLP).

Tips

Keep your laboratory records in a specific notebook. For each experiment in your laboratory notebook include:

- **title and date;**
- **objectives;**
- **reaction scheme, if applicable;**
- **risk assessment;**
- **method;**
- **results and calculations including error margins, spectra and graphs; and**
- **a conclusion.**

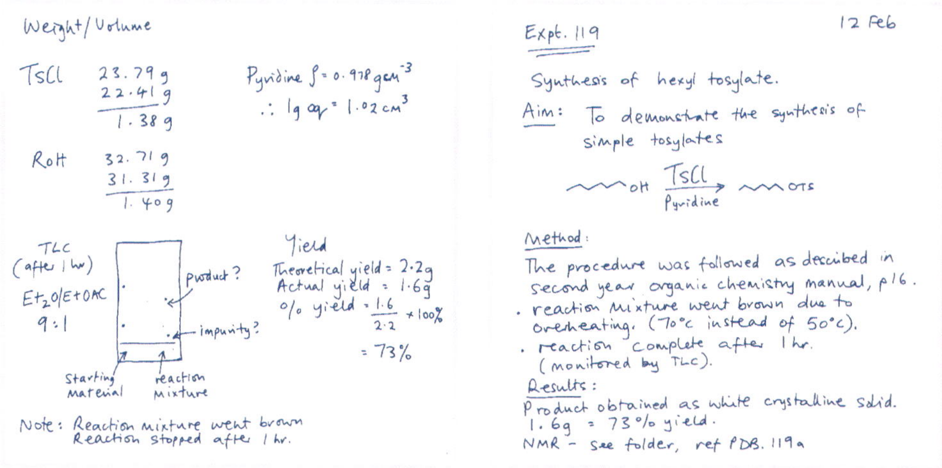
Weight/Volume

TsCl 23.79 g
22.41 g
1.38 g

RoH 32.71 g
31.31 g
1.40 g

Pyridine ρ = 0.978 gcm^{-3}
∴ 1g eq = 1.02 cm^3

TLC
(after 1 hr)
Et_2O/EtOAc
9:1
Product?
impurity?
Starting material
reaction mixture

Yield
Theoretical yield = 2.2g
Actual yield = 1.6g
% yield = 1.6/2.2 × 100%
= 73%

Note: Reaction mixture went brown
Reaction stopped after 1 hr.

Expt. 119 12 Feb

Synthesis of hexyl tosylate.

Aim: To demonstrate the synthesis of simple tosylates

OH —TsCl, Pyridine→ OTs

Method:
The procedure was followed as described in second year organic chemistry manual, p16.
- reaction mixture went brown due to overheating. (70°C instead of 50°C).
- reaction complete after 1 hr. (monitored by TLC).

Results:
Product obtained as white crystalline solid.
1.6g = 73% yield.
NMR – see folder, ref PDB. 119a

An example of a laboratory notebook is shown above.

One style is to jot data and observations on the left-hand page with a more detailed record on the right-hand page.

Make sure you understand the aims of the experiment and are aware of the hazards before you start. A risk assessment should also be completed at this time. Note the procedure in your laboratory notebook (or include a reference) and check through the timing of the experiment. Remember to include the date in your laboratory notebook.

Record your observations, especially if they differ from the procedures and include all weights and measurements if necessary. Complete the record of the experiment immediately afterwards and remember to show calculations and a conclusion.

When carrying out research the same advice applies. Include an introduction outlining what you intend to do including quantities and schemes, and reference related work. A risk assessment should also be included. File spectra and graphs and reference them in your laboratory notebook. Since the laboratory notebook is the only record of your research it is a good idea to keep a copy of all your notes, and keep this in a different place.

Remember that it may be useful to store your laboratory records electronically. In the same way as a paper-based laboratory record, you should always keep a back up copy.

Section 3
Writing a report

THIS SECTION INCLUDES ADVICE ON WRITING A REPORT WHICH COLLEAGUES MAY READ AND SHOULD BE ABLE TO UNDERSTAND

Whether you are in industry or university, a lecturer, a salesperson, a molecular biologist, an accountant or a manager you have to write reports. Report writing is a key part of most jobs. Reports inform others about recent developments in your area, include proposals for new work, or summarise presentations and discussions from a conference you attended.

Short reports (usually 4-20 pages) often don't include experimental detail, e.g. research proposals, topic reviews, project reports or a short paper. They provide an overview and it should be possible to read them quickly to identify key points. It can be difficult to summarise the information in a short report, but you should ensure that it is well presented and readable.

Full reports (usually 20-200 pages) describe the research in detail, including analysis of results and any conclusions, e.g. a full paper or thesis. When writing a research report you must select the key results from your experiments and explain the experimental design and conclusions clearly and concisely.

Short reports

Tips

- Choose a relevant title and include your name on the report;
- identify sections within your report – aim for about six sections; – e.g. Abstract, Introduction;
- structure your report using headings;
- write brief notes on each section, highlighting where you will use graphics;
- sketch out the graphics – e.g. schemes, structures, graphs; and
- use a suitable computer programme to draw chemical structures – e.g. Isis Draw or ChemDraw.

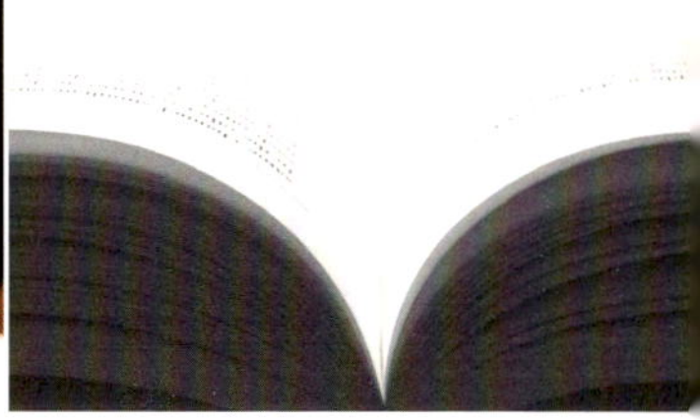

Before starting your report check the required length and format – i.e. font size and layout. Make sure you know your intended audience so you can pitch the report at the right level. Assuming you are familiar with the topic – i.e. the information you need is accessible, you have read appropriate journals/books, or have carried out a series of experiments recorded in your laboratory notebook – you can prepare to write your report.

Write a complete first draft of your report and expect to edit the content at least twice.

A short report should be:

- **clear;**
- **concise;**
- **attractively presented – i.e. with suitable graphics; and**
- **the correct length.**

Read the final report yourself and make any additional amendments. Ask colleagues for their comments.

A good abstract is essential and may be the first – and sometimes the only – section that is read in a report. Further information about writing clear and concise abstracts can be found in Section 5 of this booklet.

Full reports

Tips

A full report should include:

- **abstract or summary;**
- **introduction and background;**
- **results and discussion;**
- **experimental details; and**
- **references.**

The order of the sections may vary and can be divided up differently, but most full reports cover these areas. The main body of a full report is the results and discussion section. Your research does not have to be described in the order in which it was completed – look at the results from your research and decide the best way to explain the work.

- **Use headings and sub-headings to simplify a report;**
- **interpretation of results is usually covered in the section on results and discussion;**
- **illustrate the results with relevant schemes and diagrams;**
- **include titles for each table, scheme, graph, spectrum and figure; and**
- **start writing the introduction and experimental section at an early stage.**

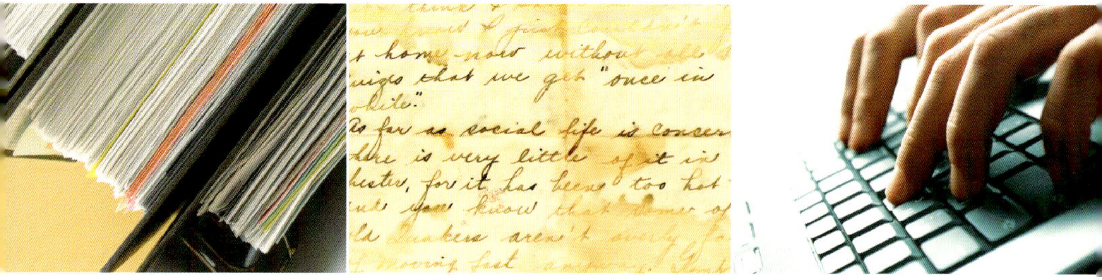

The **introduction** and **background** should contain a brief summary of the importance of the topic, a review of previous work in this particular area and the aims of the work carried out.

The **experimental** section should contain sufficient detail in order for colleagues to repeat the experiments. This section should also contain an accurate record of the data. Remember to double check the experimental data since others will use this to check their results.

A list of key **references** should be included at the end of the report.

For chemistry reports, the RSC's format for presenting experimental details is a good guide. The first issue of an RSC journal each year contains information for authors. Alternatively, guidance notes for authors are available for each journal on the RSC website, **www.rsc.org/journals**

For reports on research chemistry it is recommended that you use the same reference format as the RSC does – i.e. author's name, *journal title*, year, **volume**, page number:

J. G. Karlsson, B. Karlsson, L. I. Andersson, and I. A. Nicholls, *Analyst*, 2004, **129**, 456

For physics reports follow the Institute of Physics' format. Check out Notes for Authors on **www.iop.org/journals/nfa/index.html**

If you are writing a report on physics it is advisable to follow the reference advice given by the Institute of Physics: i.e. name(s) and initials, date published, *title of journal or book*, **volume**, and finally the page numbers:

Sharma V. K., Chattopadhyay M. K. and Roy S. B. 2007 J. Phys. D: *Appl. Phys.* **40** 1869

For biology reports see the Institute of Biology's website for information – **www.iob.org**

You should also follow the Institute of Biology's reference style: i.e. name(s), (year published), title of article, *journal title*, **volume**, (**issue**) and finally the page numbers:

MacFarlane G. R., Markwell K. W. and Date-Huxtable E. M., (2006), Modelling the research process as a deep learning strategy, *Journal of Biological Education*, **41(1)**, 13-20.

Proof-reading

All reports, posters and slides should be proof-read before presentation. Always ask a colleague to proof-read your work – it is often difficult to spot mistakes in your own work.

Tips

Check the following:

- **spelling;**
- **grammar;**
- **style; that**
- **tables, figures, spectra, are numbered correctly and consecutively; and that**
- **figures and illustrations are labelled properly.**

It is also advisable to proof-read for no more than 20 minutes at a time. You may have to read the report more than once – first check for spelling, grammar and style; and then check the content.

Do not rely on the computer spell-check to proof-read your work especially if you have included scientific terms. Spell-checkers are not fool-proof.

Notes

Section 4
Information retrieval

THIS SECTION BRIEFLY DESCRIBES SEARCH PROCEDURES WHICH ARE WIDELY USED BY SCIENTISTS - DATABASES, ABSTRACTS AND THE INTERNET

Information retrieval is a key part of many jobs – e.g. to supplement lecture notes, to collect safety information for a risk assessment, or to find out about recent developments in a particular area.

Scientific information is available in books, journals and databases. Information searches were once entirely paper based, but information is now usually available through computer databases and the internet, and historic information is also being made available online. This simplifies the search process considerably.

Most universities offer training to students on information retrieval. Check the library for training sessions on using databases. Librarians can also assist you with paper-based searches. Members of the RSC can use the facilities of the RSC Library and Information Centre at Burlington House, this can also be accessed online at www.rsc.org/library.

Abstracts and Databases

Abstracts are available for most research fields in paper-based and electronic versions. There is a range of online databases available which enable you to search the literature and keep up to date with current research.

Chemical Abstracts

Chemical Abstracts is a comprehensive database of most chemistry publications and contains more than 27 million abstracts. It provides abstracts of articles in journals, books, conference proceedings, patents and reports in both paper-based and electronic (CD-ROM) versions. Details about the electronic version can be found at www.cas.org. Most libraries have the paper-based version of Chemical Abstracts as well as electronic access.

Each volume covers a six-month period, and there are a set of indexes available for each volume. These indexes enable you to search by subject, author, chemical substance, formula, patent and by CA registry number. The indexes give the entry numbers for abstracts in that volume. There are collective indexes for each five-year period.

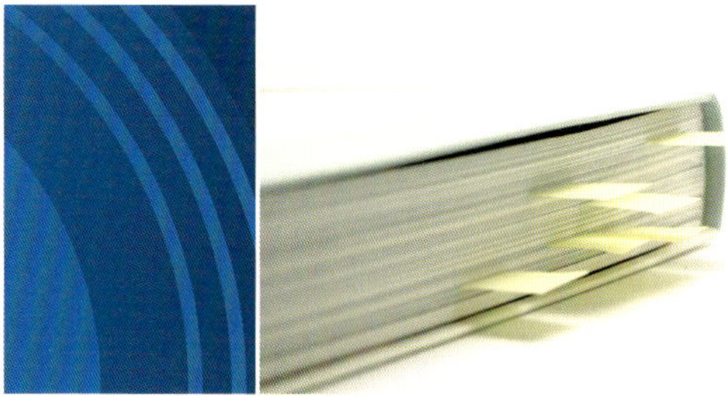

author

Bailey, Patrick D. See Ridyard, Colin H.

–; Higgins, Stanley D.; Ridyard, Colin H.; Roberts, Stanley M.; Rosaire, Georgina M.; Whittaker, Roger A.; Willets, Andrew J.

Selective biohydroxylation of 1-substituted adamantanes using Absidia cylindrospora (I.M.I 342950), 162964u

–; Morgan Keith M.

A total asymmetric synthesis of (-)-suaveoline, 143105y

subject & entry number

Biological Abstracts

Biological Abstracts is the largest printed reference publication for life, biomedical and environmental sciences. The electronic database can be accessed via the ISI Web of Science. Thousands of life science journals, international meetings and books from more than 90 countries are monitored for inclusion. The database contains bibliographic information from 1926 to present day ranging from traditional areas of biology to related fields such as pharmacology and bioengineering.

Look up a particular index to get the entry number. To locate the abstract find the entry number in the volumes of chemical abstracts. For chemicals it is advisable to use the formula index.

volume entry number

125: **143105y A total asymmetric synthesis of (-)-suaveoline.** Bailey, Patrick D.; Morgan, Keith M. (Dep. Chem., Heriot-Watt Univ., Edinburgh, UK EH14 4AS). *Chem. Commun. (Cambridge)* **1996,** (12), 1479-1480 (Eng). A total synthesis of (-)-suaveoline (**I**) from...

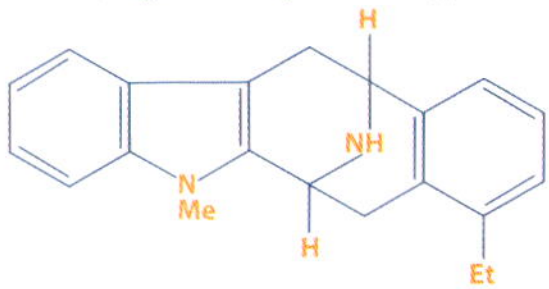

Physics Abstracts

Physics Abstracts is the world's major guide to recently-published research in all areas of physics including particle, fluid and solid-state, bio-and astro-physics. Around 180 000 abstracts are published each year in 24 issues. Each issue includes a subject guide and indexes. Cumulated indexes are published twice a year and include subject, author and other indexes.

INSPEC **(http://edina.ed.ac.uk/inspec/)** is the premier database for access to the world's leading scientific and technical literature in physics, electrical engineering, and associated areas. INSPEC holds over 5.75 million abstracts from scientific and technical journals, conferences and other publications. It is widely available in higher education institutions.

Please note that Chemical Abstracts is retrospective i.e. volume 146 (January - June 2007) mainly covers papers published in late 2006.

Web of Science

Web of Science forms part of the ISI (Institute of Scientific Information) Web of Knowledge. It is a collective reference to the electronic publications of ISI, giving information on articles published in several thousand journals worldwide. A username and password are required to access this resource however, most HE institutions subscribe to the service. Access to the Web of Knowledge is available online at **http://wok.mimas.ac.uk/**

Typically databases can be searched for articles of interest by using words or phrases in article titles, keywords or abstracts, by author names or addresses, by journal names or other identifying information.

Web of Science provides accurate references, titles of papers, authors names and addresses, keywords, and in most cases abstracts, for all papers included in the Science Citation Index. The search results can be printed, downloaded and e-mailed. Web of Science Proceedings indexes the published literature of the most significant conferences, symposia, seminars, colloquia, workshops and conventions in a wide range of disciplines in science and technology.

Using a combination of descriptors – e.g. author(s), keyword(s) or year of publication, the computer searches the database and provides a list of references that match the search criteria. The references can be displayed either as a full record, or more concisely by title, author and source.

Searches can be confined to:

- **words in title, keywords, or abstracts;**
- **authors' names and/or addresses; enter an author's name as: "Surname Initials";**
- **journal names;**
- **references to earlier work (citation searching); and**
- **specific years (searching by date refers to the date of addition to database).**

Using the following operators is useful during searches:

*	**e.g. optic* gives matches for optic, optics, optical**
And	**e.g. laser and cooling in the title search matches papers with both words in the title**
?	**e.g. alk?ne gives matches for alkane, alkene, alkyne**
Or	**e.g. phosphoric or phosphonic searches for papers containing either of these words.**

Internet

There are numerous useful websites for scientists including professional bodies, universities, industrial companies, recruitment agencies, journals and newspapers. It is often possible to register for conferences online as well as participating in online science discussions.

Useful points of contact include:

- **www.rsc.org** (Royal Society of Chemistry) and the ChemSoc network - **www.chemsoc.org**;
- **www.iop.org** (Institute of Physics) and for physics global news and information **www.physicsweb.org**;
- **www.iob.org** (Institute of Biology);
- **www.epsrc.ac.uk**;
- **www.scitech.ac.uk**; and
- **www.bbsrc.ac.uk**.

The following sites select and list a range of sites, which are considered to be authoritative, factually correct and up-to-date:

www.bubl.ac.uk/link - provides access to selected internet resources covering all academic subject areas;

www.hero.ac.uk - an internet portal for academic research and higher education in the UK;

www.intute.ac.uk/healthandlifesciences/medicine - a gateway to high quality biomedical internet resources.

Electronic journals

Many journals are now published in electronic format. In general, free access to the full text of the electronic version is only available if the library subscribes to the printed version of the journal, or pays a separate subscription. Some journals that are newly available in electronic form may be available free of charge for a limited period. Check out the journal publisher's internet site for more information. Another useful resource is the RSC journal archive, this provides access to more than 160 years of RSC journal articles and is available to subscribers at **www.rsc.org/publishing**

The internet can help you keep abreast of developments, to do this effectively and efficiently, you should:

- **bookmark key internet sites**
- **use search engines and key words to sort through the web periodically**

Information is available from a wide range of sources, remember to use your library services to help you access everything you need during your studies and research.

Section 5
Preparing an abstract

THIS SECTION GIVES GUIDANCE ON WRITING ABSTRACTS FOR BOTH POSTER AND ORAL PRESENTATIONS AND WRITTEN REPORTS

If you decide to prepare a poster or give a presentation at a conference or write a report you will be asked to submit an abstract summarising your work. Guidelines for abstracts vary with every conference, however some general rules apply.

Make sure your title clearly describes the content of your abstract; try not to be too cryptic! In general your title will need to be centre-aligned and appear in bold, lowercase letters. Authors and affiliations should appear below the title in italics with the main author's name underlined and the presenting author marked with an asterisk (*). The order in which author names are written can be a contentious issue. Check with your supervisor and the conference guidelines, but if in doubt put them in alphabetical order.

Your abstract should be no more than one side of A4 with single-spaced text. In general the font and font size will be defined by your institution or the conference organisers but if this is not stated it is best to choose a clear, easy-to-read font, such as Arial or Times New Roman, in not less than size 12. Text should be left-aligned, making sure you adhere to the conference margin guidelines. Abstracts are often bound to form a conference booklet and set margins are needed in order to ensure that no text is obscured in the binding process.

Including pictures and diagrams helps break up the text and improves the look of an abstract. Also, pictures can often explain topics which may be difficult to define in words without a long and convoluted description.

You may need to reference other people's work during your introduction, try not to include more than five references and make sure you adhere to the conference guidelines for numbering and style.

What is a green solvent? A comprehensive framework for the environmental assessment of solvents

Christian Capello, Ulrich Fischer * and Konrad Hungerbühler

ETH Zurich, Institute for Chemical and Bioengineering, Wolfgang-Pauli-Strasse 10, 8093 Zurich, Switzerland.
E-mail: ufischer@chem.ethz.ch; Fax: +41-44-632-1189; Tel: +41-44-632-5668

Received 1st December 2006, Accepted 22nd February 2007
First published on the web 9th March 2007

Solvents define a major part of the environmental performance of processes in chemical industry and also impact on cost, safety and health issues. The idea of "green" solvents expresses the goal to minimize the environmental impact resulting from the use of solvents in chemical production. Here the question is raised of how to measure how "green" a solvent is. We propose a comprehensive framework for the environmental assessment of solvents that covers major aspects of the environmental performance of solvents in chemical production, as well as important health and safety issues. The framework combines the assessment of substance-specific hazards with the quantification of emissions and resource use over the full life-cycle of a solvent. The proposed framework is demonstrated on 26 organic solvents. Results show that simple alcohols (methanol, ethanol) or alkanes (heptane, hexane) are environmentally preferable solvents, whereas the use of dioxane, acetonitrile, acids, formaldehyde, and tetrahydrofuran is not recommendable from an environmental perspective. Additionally, a case study is presented in which the framework is applied for the assessment of various alcohol–water or pure alcohol mixtures used for solvolysis of p-methoxybenzoyl chloride. The results of this case study indicate that methanol–water or ethanol–water mixtures are environmentally favourable compared to pure alcohol or propanol–water mixtures. The two applications demonstrate that the presented framework is a useful instrument to select green solvents or environmentally sound solvent mixtures for processes in chemical industry. The same framework can also be used for a comprehensive assessment of new solvent technologies as soon as the present lack of data can be overcome.

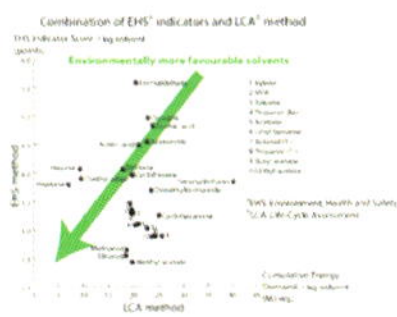

An example of an abstract is shown above.

Acronyms are often used within the chemical community, however people outside your field may not understand them. To avoid confusion try not to use acronyms in your abstract, unless it is not possible to write what you need in any other format.

Remember, your abstract will encourage people to read your report, visit your poster or attend your presentation; it is down to you to sell your work and yourself in a single page of text. Be concise and to the point providing an introduction to your work. Try to intrigue the reader and make them want to find out more.

Section 6
Giving a presentation

THIS SECTION INCLUDES IDEAS TO HELP YOU GIVE EFFECTIVE PRESENTATIONS USING POWERPOINT® SLIDES AND A DATA PROJECTOR

Scientists are required to give oral presentations throughout their careers. Giving a formal talk can be a daunting task, even if the content is well prepared and the presentation thoroughly rehearsed.

Selecting the correct material, keeping to the time allocated, pitching the content at the right level for the audience, using visual aids and producing informative handouts are all essential for a successful presentation. When giving a presentation, be confident – remember that, in general, you know more about the topic than your audience. Plan a presentation that suits your style and is clear and interesting for the audience. Avoid jargon and abbreviations wherever possible.

Visual aids

There are numerous visual aids for presentations – e.g. a blackboard, whiteboard, flipchart, video, DVD/CD-ROM, data projector. In general, presenters use data projectors at conferences, although several visual aids can be used during a presentation. If you are using a PowerPoint® presentation it is worthwhile having a set of overheads as a back-up. Providing handouts for your audience is often welcome, printing out three slides per page with space to make notes ensures the audience doesn't have to copy the content of the slides during your talk.

Tips

A presentation should be:

- **suitable for the audience;**
- **informative;**
- **interesting; and**
- **the appropriate length.**

When preparing a presentation, divide the content into three sections – e.g. introduction, core, and summary. The "core" should be the longest section of the presentation, and a summary of the main points should be given in the final few minutes.

Sketch out what you want to include in the presentation use short headings, bullet points, schemes/tables and graphs/structures.

When planning slides keep them simple - use a large font size and appropriate colour (for example avoid using yellow text on a white background) and use diagrams. Each overhead/slide should be displayed long enough for the audience to be able to use the information, or to act as reminders of key points. Prepare high quality slides using a suitable computer program, such as PowerPoint®.

Check that everything on a slide can be read from a distance. Do not scan intricate text/diagrams from a book and insert into a slide as the audience will not be able to read them. Number your slides and ensure that they are in the correct order.

Practise your presentation several times in advance. Ask a friend or colleague to listen to the presentation and to give some feedback. If necessary, write out the content of the presentation in full. However, do not read it out during the actual presentation – prepare headings on cards or use the content of your slides as a prompt.

On the day of the presentation check that the computer and data projector are working and focused, that you know how to operate them, and that the images are visible from the back of the room. Speak slowly and clearly remembering that the audience is not as familiar with the topic as you are. Make sure that you are audible from the back of the room. Look at the audience during the presentation, refer to the overheads/slides occasionally and remember to try to avoid blocking the audience view of the screen.

Finish the presentation with a summary and thank the audience for their attention. You are often required to answer questions – don't rush replies and if you don't know the answer say so.

Afterwards, it is a good idea to ask for feedback from colleagues so you can improve your technique in future. Listen to colleagues' presentations and note good practice to help you with future presentations.

Section 7
Preparing a poster

THIS SECTION GIVES ADVICE AND TIPS ON PREPARING INFORMATIVE POSTERS, WHICH WILL ALSO CATCH THE READER'S ATTENTION

Posters are increasingly used for presenting scientific information at meetings and conferences. There are various styles for presenting your work as a poster. It is essential that whichever style you choose the poster should be eye-catching, legible and contain enough information to give the reader an overview of the work, without being overcrowded. Most conference organisers specify the size of posters and poster headings.

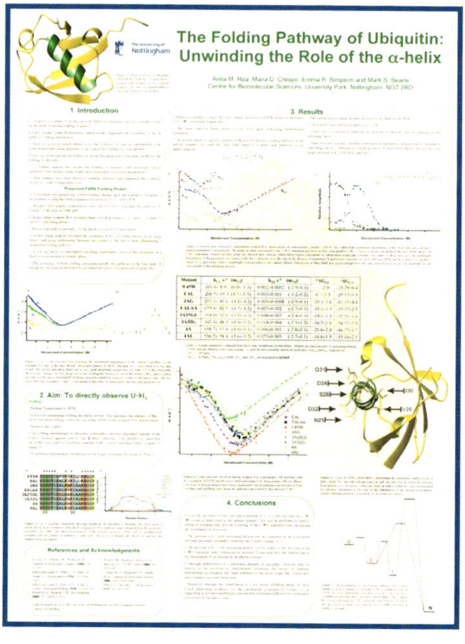

Conference organisers often request an abstract of your work before selecting posters for exhibition. The abstract should include a summary of your work and your contact details – for more information about writing abstracts turn to Section 5 of this booklet.

Tips

A poster should:

- **have visual impact;**
- **have a clear layout;**
- **be readable in 3-4 minutes; and**
- **be legible from two metres away.**

Before preparing a poster check the size required and the intended audience. Compile a list of headings related to the work you are presenting and identify structures, tables, spectra or pictures you intend to include. Prepare draft text according to the list of headings and sketch schemes, figures and graphs. Check the length of the draft and edit if necessary, before preparing the final version of the poster.

The poster should contain a brief background to the work, results and discussion, and a conclusion. It should also contain a list of references and acknowledgements. A poster should highlight the key points of your work; make sure you don't just reproduce pages of text from reports.

Remember to keep within the size limit for the overall poster and check whether it needs to be portrait or landscape in design.

SUBSTITUTED FERROCENES AS LIQUID CRYSTALS Nick Bolton, Sheffshire College	
1. AIMS	2. BACKGROUND TO RESEARCH
3. TARGET STRUCTURES	4. SYNTHESIS
6. OTHER DATA	5. SPECTRA
7. CONCLUSIONS	8. REFERENCES 9. ACKNOWLEDGEMENTS

It is worthwhile having a handout to accompany your poster, including contact details and an abstract of your work. This can contain more detailed information than the poster.

Posters should be self-explanatory even in the absence of the author. Use simple figures and schemes and add colours for emphasis. Having a coloured background adds impact, however make sure this does not obscure the information you are trying to convey.

Laminating your poster creates a professional look but is not vital.

Ensure that the text is readable from a distance of at least two metres – use at least 18 point font size for main text and 32 point size for headings.

The final layout should be easy to follow – number each section or use arrows to ensure that the content flows. Boxes are also effective to highlight schemes and spectra.

Effective posters are difficult to produce – they must catch the reader's attention, retain their interest and get the message across.

When you are participating in, or attending, a poster session it is useful to look at other posters and note any good ideas. Try to include some of these ideas the next time you are preparing a poster.

Section 8
Meetings and networking

THIS SECTION GIVES ADVICE ON THE ROLE OF EACH PERSON ATTENDING A MEETING. IT ALSO INCLUDES SOME ADVICE ON NETWORKING

Meetings are a key part of most jobs. Successful meetings depend on preparing a concise agenda, structured discussion, and producing a clear set of minutes or actions afterwards. Meetings are a good way to network with others - it is important to establish contacts with others, both within and outside your department.

Tips

- **The secretary should consult the chairperson and draft an agenda for the meeting;**
- **the agenda should be circulated in advance;**
- **the chairperson should direct the discussion; and**
- **all decisions should be followed up afterwards.**

The chairperson should plan a rough timetable for the meeting and allow sufficient time for discussion of each item on the agenda. At the end of each item, it is important that the chairperson summarises specific decisions and identifies any actions, when they should be taken and who should carry them out.

The secretary's role is to take minutes of the meeting. This involves listening to the discussion and noting important points, particularly conclusions, decisions and action points (as well as who will do them). Highlight any items that need to be clarified with someone after the meeting. Minutes should be written in reported speech - e.g. 'It was agreed that John Smith would produce a poster', not 'John Smith said that he would produce a poster'.

Write the minutes as soon as possible after the meeting – the minutes should be accurate, clear and concise. They are not a verbatim record of what was said at the meeting, just the key points and actions. Ask the chairperson to check the accuracy and content of the minutes before sending them to those who attended the meeting.

Committee members attending a meeting should read the agenda beforehand and note any points they would like to make during the meeting. Attract the chair's attention if you have a point to make in the discussion. Remember to read the minutes when you receive them and note any actions you are required to take forward.

Networking

Whether you are carrying out research in a university or working in industry you need to network. Try to develop a means of networking with colleagues outside your own team or department, and when attending conferences establish new contacts with other delegates.

Tips

- **Take the opportunity to talk with colleagues informally (i.e. during coffee breaks) regarding work and career issues;**
- **join electronic mailboxes in relevant areas;**
- **at conferences, participate in poster and question sessions; and**
- **take opportunities to demonstrate you have good ideas e.g. during committees or decision-making processes.**

Networking is key to your career development, try to develop this skill with colleagues within and outside of your department as well as at conferences and meetings.

Section 9
Time management

THIS SECTION INCLUDES TIPS ON MANAGING YOUR TIME EFFECTIVELY

Managing your time and your work is vital whether you are studying, working in industry or carrying out postgraduate research. You need to plan your work in the short term, but you also need to set some long term objectives.

Time management is central to your work. You have to meet deadlines and achieve aims within a given timescale. There may be some factors which prevent you from achieving your aims in the time available to you such as lack of clear objectives, lack of effective planning, poor communications, and too many interruptions.

You may have several projects to work on with similar deadlines. Prioritising your workload is essential. Employers need employees who can prioritise and meet deadlines. Set long, medium, and short-term priorities. Remember not to plan your time so tightly that an unanticipated pressure ruins your planning. It is a good idea to make allowances for unexpected pressures when planning your work.

It is important for you to identify factors that might obstruct your progress. You then need to plan how you are going to manage your time effectively. A list of suggestions is given below. Some of these suggestions may seem obvious, but it is necessary to 'stand back' from your work occasionally and assess how you manage it.

Prepare and organise your work:

- **keep your desk tidy;**
- **be conscious of time;**
- **compile a daily list of tasks;**
- **set realistic deadlines;**
- **try not to take on too much; and**
- **finish one task before you go on to the next one.**

Organising or chairing meetings requires good time management. Plan meetings well in advance. If you are chairing the meeting allocate a period of time to each item on the agenda and keep to this timing during the meeting. Remember that meetings should have a clear objective and you should not discuss other topics during the meeting.

Meetings:

- **if you are organising a meeting, make sure you have clear objectives, start and end times;**
- **do not allow meetings to be interrupted; and**
- **when setting up, or attending a meeting do not waste other people's time.**

Short and long term planning is key to both studying effectively and your long term career goals. Try to develop this throughout your studies.

Notes